Microlearning

For

Educators

Intentional Strategies to Transform Professional Development in the Digital Age

Terry Ross, Ph. D.

ISBN: 978-1-7367989-2-8

Table of Contents

Introduction

Microlearning for Educators: Intentional Strategies to Transform Professional Development in the Digital Age" is designed as a guide for the new generation of teachers—those who find themselves in the unique position of teaching while learning, adapting while instructing, and growing while guiding. New Teachers are entering education in an era where the pace of change is relentless and the demand for continuous professional growth is more critical than ever. With "The Great Resignation" having an impact on the classroom, the concept of microlearning emerges not just as a method but as a necessity for educators striving to stay ahead. Millennials make up a significant portion of the teaching force. We also have a huge number of second-career educators coming into the classroom. Many of our New Teachers are from retail, corporate, and manufacturing. The challenge of preparing an intergenerational teaching force for the future while honoring the wisdom of the past has never been more daunting. As educators, we are tasked with not only imparting knowledge but also embodying the adaptability required to navigate the communities where we teach and learn.

Education, as we know it, is undergoing a transformation, influenced heavily by the influx of millennial educators into the workforce. This generation of teachers brings with them not just the promise of fresh perspectives and energy but also unique challenges and learning preferences. They are **digital natives**, accustomed to rapidly consuming information through the internet, social media, and, notably, podcasts. Their entry into the educational field signals a shift towards more dynamic, engaging, and flexible teaching and learning methodologies. This shift is also needed. We need teachers that can relate to the students that they teach. Administrators must shift how they view teaching and learning, as well as how they provide support to teachers.

For years, the approach to professional development within education has been characterized by extensive, time-consuming sessions that, while often rich in content, do not necessarily align with the realities of a teacher's daily routine or the rapidly changing educational environment. In an era marked by information overload and constant demands on our attention, the traditional models of professional growth have shown their limits. It is time for a paradigm shift—a move towards a more focused, efficient, and impactful form of learning. This book introduces the concept of microlearning as a powerful alternative to traditional professional development methods, tailored to meet the needs of today's intergenerational teaching force.

Microlearning, the practice of delivering content in small, manageable segments, aligns perfectly with the needs and preferences of these educators. It offers a solution to the often-overwhelming challenge of professional development, breaking down complex concepts into digestible pieces that can be easily understood and applied. This approach not only facilitates immediate application in the classroom but also supports the millennial educator's lifestyle, allowing for learning that is both time-efficient and impactful.

This book aims to explore the multifaceted benefits of microlearning for professional development. It delves into how microlearning can be effectively integrated into teaching practices, offering a roadmap for new teachers to enhance their skills, engage their students, and ultimately, transform their classrooms. Through a blend of theoretical insights and practical strategies, it serves as a comprehensive guide for educators looking to harness the power of microlearning to fuel their professional growth and adapt to the evolving demands of the education sector.

As we journey through the chapters, we will uncover the principles behind effective microlearning, explore the tools and technologies that support it, and provide actionable advice on creating a personalized microlearning plan. This book is not just about understanding microlearning; it's about experiencing its transformative potential firsthand.

"Microlearning for Educators: Intentional Strategies to Transform Professional Development in the Digital Age" is not just a critique of the status quo; it is a roadmap to a more effective and sustainable model of teacher development. It advocates for a shift from the 'more is better' mindset to a 'less is more' philosophy, where quality trumps quantity, and learning is integrated seamlessly into the fabric of daily professional life. By adopting microlearning principles, schools can create a culture of continuous, incremental growth that respects the time and diverse needs of all educators, from the digital natives to the seasoned veterans.

As we embark on this journey together, my hope is that this book will serve as both a mirror and a window—a mirror reflecting the shared challenges we face in education, and a window to new possibilities for nurturing a teaching force that is not only competent but also resilient, innovative, and prepared for the future. Welcome to "Micro Dosing Professional Development," where we embrace the power of doing less to achieve more.

Chapter 1: The Rise of Microlearning in Professional Development

Professional Development in Education can be both rewarding and exhausting. For years, teachers have dreaded going to professional development because it was always one-size-fit-all, long, and not related to what each content or individual teacher needed. The concept of microlearning has emerged as a powerful approach to professional development that aligns perfectly with the modern educator's need for flexible, efficient, and impactful learning. Professional development in education is undergoing a profound transformation, characterized by a shift towards more agile and learner-centered approaches. At the forefront of this evolution is microlearning, a methodology that promises to redefine how educators grow professionally, adapt to new teaching challenges, and meet the ever-changing needs of their students. As educational demands evolve, so must the strategies educators employ to enhance their skills and teaching practices. This chapter explores the concept of microlearning, its growing relevance in education, and its myriad benefits to educators committed to lifelong learning and improvement.

Understanding Microlearning: A Primer

Microlearning breaks down complex information into small, manageable units, making learning more digestible, engaging, and applicable. Typically, these learning units are short (ranging from a few seconds to up to 20 minutes), focused on a single objective, and designed to be accessed at the learner's convenience. This approach contrasts with traditional professional development models, which often involve longer, more generalized sessions that can be overwhelming and less immediately applicable.

New Teachers are overwhelmed with all of the newness of the job and role of classroom teacher. When New Teachers arrive at their new school, they are given:

- content/subject area
- their classroom (THE KEYS)
- class roster
- curriculum guides
- pacing guides
- school-wide discipline plan
- emergency management plan
- assessment calendar
- activity calendar
- student demographic data (academic and climate data)
- PLC Schedule
- Supplies

This is just the starter package for New Teachers. The list grows by the end of the first quarter. Repetition and Microlearning capitalizes on the cognitive benefits of processing smaller pieces of information, which can lead to higher retention rates and more effective application of new knowledge. It acknowledges the modern educator's busy schedule and varied professional learning needs, offering a flexible solution that fits learning into the crevices of a packed day without sacrificing depth or quality.

The Shift Towards Microlearning in Education

The emergence microlearning in professional development is not coincidental but a response to several broader shifts within education and society at large. The digital age has transformed how we access and interact with information, favoring quick, on-demand knowledge over lengthy, static learning experiences. Educators, like their students, now navigate a world where information is abundant, attention spans are shorter, and the ability to quickly update and apply new skills is paramount.

Additionally, the entry of the millennial generation into the teaching workforce has accelerated the adoption of microlearning. Accustomed to the immediacy of digital information and communication, millennial educators often prefer learning opportunities that are not only flexible and diverse but also aligned with their digital fluency. This demographic shift has prompted educational institutions and professional development providers to rethink their approaches, moving towards microlearning as a way to cater to these evolving preferences and needs. The influx of second-career educators with backgrounds in retail, manufacturing, military, and entrepreneurs has increased the need for more on-demand learning opportunities. Thirty years ago, most educators majored in education. That is not the norm anymore. The dire need for teachers has pushed districts and schools alike to hire willing people with a degree. The key word is 'willing' people, we often believe that if you have the will, we will teach you the skill.

Advantages of Microlearning for Continuous Professional Growth

Microlearning offers a plethora of advantages for educators seeking to enhance their professional skills and teaching effectiveness continuously. Here are some of the key benefits:

- **Flexibility and Accessibility:** Microlearning allows educators to learn at their own pace and according to their own schedules, making it easier to fit professional development into a busy teaching calendar.

- **Immediate Applicability:** The focused nature of microlearning content means that educators can quickly apply what they've learned to their teaching practice, seeing immediate benefits in the classroom.

- **Enhanced Retention:** Studies suggest that learners are more likely to retain information presented in short, focused segments. This enhanced retention is crucial for educators who must keep abreast of best practices and evolving pedagogical theories.

- **Personalization:** Microlearning can be easily tailored to individual learning needs and interests, allowing educators to focus on areas where they want to grow or need additional support.

- **Cost-Effectiveness:** By utilizing digital platforms and resources, microlearning can be a cost-effective way for educational institutions to provide ongoing professional development without the need for extensive physical resources or time away from the classroom.

As we delve deeper into the nuances of microlearning and its application in the following chapters, it becomes clear that this approach is not just a temporary trend but a significant shift towards more dynamic, efficient, and effective professional development for educators. The emergence of microlearning reflects a broader understanding that, in the fast-paced world of education, the methods by which we learn and grow professionally must evolve to keep pace with the demands of teaching in the 21st century.

Chapter 2: Microlearning as a Tool for New Teachers

By the time New Teachers arrive on the first day of school, they are already overwhelmed with systems and processes. In the journey of a new teacher, the initial days are pivotal. They are filled with the excitement of shaping young minds, coupled with challenges that test resilience, adaptability, and the capacity to grow. Microlearning emerges as an indispensable tool for new teachers, offering strategies that bridge the gap between theoretical knowledge and practical application, enhance classroom practice, and allow for the customization of learning paths. Observation from my extensive experience in education and in leadership development, I've witnessed firsthand the transformative power of microlearning in addressing early career challenges, enhancing teaching methodologies, and fostering personalized professional growth. New Teachers are overwhelmed with so much information during their onboarding, along with the policy manual and Teacher Handbook, that they don't know where and what to prioritize. It is common for New Teachers to get lost in the weeds of all the new information.

Bridging the Gap: Microlearning for Early Career Challenges

A steep learning curve often marks the transition from educational theory to classroom practice. New teachers are expected to manage diverse classrooms, design effective lesson plans, engage students with varying learning needs, and much more. Here, microlearning can play a crucial role. It offers just-in-time learning that helps new educators tackle immediate challenges without feeling overwhelmed. For instance, during my tenure as the principal of Getwell Elementary School, Bennett High School, and later at Kingsbury High School, we implemented microlearning strategies to support our teachers in areas like classroom management, establishing a growth mindset, and differentiated instruction. These focused, bite-sized learning opportunities empowered them to address specific issues as they arose, facilitating a smoother transition into their roles and fostering a supportive learning environment for their students.

We established our 'big rocks' or 'focus areas' at the beginning of the year. Our School Improvement Plan was a real living document that we used. We started the school year with Professional Development (PD) on the instructional strategies that we were going to implement for the entire year. All the PD, articles, case studies, and any other data and resources we needed were given to all the teachers in a notebook. We followed the game plan and implemented the strategies that we outlined as an Instructional Leadership Team. From that notebook, we launched our microlearning strategies for the entire year. We were usually finished implementing our strategies by October. We spent the remainder of the year rotating microlearning opportunities on all the strategies so that the teachers could remember our focus areas, learn different ways of implementing the strategies, and see different peers present how they were using the strategies. It was not a one-time PD at the beginning of the year. Again, all of the focus strategies were introduced at the beginning of the school year and for the remainder of the year, we refined our strategies and added new ones to our toolbox as needed.

Enhancing Classroom Practice with Bite-Sized Learning

Microlearning isn't just about overcoming challenges; it's also about elevating the teaching and learning experience. New teachers can incrementally enhance their classroom practice by breaking down complex educational strategies into manageable segments. This approach aligns with the cognitive science principle that smaller chunks of information are easier to understand, retain, and apply. In my experience, providing teachers with short, targeted professional development sessions on topics such as engaging teaching methodologies or the integration of technology in the classroom led to noticeable improvements in student engagement and academic outcomes. These microlearning sessions, often delivered through formats like video tutorials, podcasts, interactive quizzes, or quick-read guides, allowed teachers to quickly integrate new ideas and techniques into their teaching, making their classroom practice more dynamic and effective.

I was the King of quick reference guides. I think the military part of me made me create a quick reference guide for everything we implemented. The teachers could always look at a step-by-step one-pager for every strategy we implemented.

Customizing Learning Paths with Microlearning Modules

One of the most significant advantages of microlearning is its ability to cater to individual learning needs. New Teachers come from diverse educational backgrounds and possess varied sets of skills and knowledge. Microlearning modules can be tailored to meet these unique needs, allowing educators to focus on areas where they require further development. During my leadership roles, both at the district level and within individual schools, I advocated for the use of microlearning platforms that offered a wide range of modules covering different aspects of teaching and learning. This approach enabled our teachers to personalize their professional development, choosing content that matched their interests and areas for growth. As a result, they felt more empowered and confident in their roles, which, in turn, positively impacted student achievement and school culture.

Some new teachers will be great with content but need support with classroom management, and others will have great relationships with the students but need support with their perspective content area. This is where customizing each teacher's learning path is essential. We can give each teacher what they need without interrupting the learning opportunities for everyone else. It also empowers the teacher when we give them what they need to be successful. It gives them agency over their own learning.

Microlearning serves as a powerful tool for new teachers, offering solutions that bridge the gap between theory and practice, enhance classroom effectiveness, and provide opportunities for personalized professional development. Through the strategic implementation of microlearning, educators can navigate the complexities of their early years in the profession with greater ease and confidence, setting the stage for a rewarding and impactful teaching career. This is how microlearning supports teacher retention.

Chapter 3: Millennial Teachers and Their Learning Preferences

The teaching profession is experiencing a significant demographic shift as millennials become the predominant generation within the workforce. This new generation of teachers brings with them distinct preferences for how they learn and engage with professional development. Leveraging my extensive experience in educational leadership, particularly in roles that demanded innovative training and development strategies, I have observed the impact of tailoring professional development to align with millennials and their learning preferences. This chapter explores the characteristics of millennial teachers, their affinity for digital learning formats like podcasts and microlearning, and how integrating these modern learning formats can enhance professional development initiatives.

Millennials in the Teaching Profession: A Demographic Shift

Millennials born between 1981 and 1996, are now the largest generation in the U.S. labor force, and the teaching profession is no exception to this trend. As this generation steps into educational roles, their preferences and expectations are shaping the future of professional development in education. Research indicates that millennial teachers value flexibility, technology integration, and opportunities for collaborative learning (Mehta & Aguilera, 2020). Recognizing these preferences is crucial for designing professional development programs that not only engage millennial educators but also effectively contribute to their growth and retention in the profession.

We all win when we meet our teachers where they are and engage them in their growth and development process. The biggest winners are our children. Everyone wants to know that they are seen and that they matter. One of the biggest ways to let New Teachers know that they matter is by giving them a choice in their learning and offering explicit feedback with support and encouragement.

Podcasts and Short Snippets: Catering to Millennial Learning Styles

Millennials' propensity for digital media and on-the-go learning has made podcasts and short learning snippets particularly appealing. These formats cater to millennials' desire for flexibility and efficiency, allowing them to engage with professional development content during commutes, breaks, or other moments throughout their day. Research by Johnson & Bartlett (2019) found that podcasts and similar audio-based learning tools can significantly enhance learning outcomes for educators, providing a convenient and engaging way to access the latest educational theories, classroom strategies, and inspirational stories from seasoned professionals. During my time with Shelby County Schools, we implemented a series of podcast-based professional development sessions that allowed teachers to explore diverse educational topics at their own pace. The feedback was overwhelmingly positive, with many educators highlighting the ease of integrating these learning opportunities into their busy schedules.

It was Millennials who introduced me to listening to podcasts. They would text me podcasts to listen to, and I began to learn from the podcast and I began to like them. YouTube became all our friends when we were ready to look for a new strategy or how to teach a new standard. We have always used technology as a means of learning opportunities. With microlearning, we can be more intentional with our support for New Teachers.

Integrating Modern Learning Formats into Professional Development

The integration of modern learning formats such as microlearning, social media, and virtual communities of practice into professional development programs can significantly enhance the learning experience for millennial teachers. These formats support personalized learning paths, enable immediate application of new knowledge, and foster a sense of connection and community among educators. For example, in my role as a consultant and trainer with various educational organizations, I advocated for the use of interactive online platforms that offered microlearning modules tailored to specific teaching skills and strategies. The pandemic provided us with the urgency to begin to explore more digital platforms for professional development. This approach not only accommodated individual learning preferences but also encouraged a culture of continuous, self-directed learning. Furthermore, integrating these modern learning formats into professional development aligns with the findings of Garet et al. (2011), who emphasize the importance of active learning, collaboration, and the application of knowledge to practice in effective professional development.

Recognizing and catering to the learning preferences of millennial teachers is essential for developing and delivering effective professional development programs. By embracing digital learning formats such as podcasts, microlearning, and social media, educational leaders can provide engaging, flexible, and impactful learning experiences that resonate with this new generation of educators. As we continue to navigate the shifting demographics of the teaching profession, the successful integration of these modern learning formats will be vital to fostering a well-prepared, motivated, and innovative teaching workforce.

Chapter 4: Microlearning Strategies for Effective Teaching

As educators, our mission extends beyond imparting knowledge; it encompasses creating environments where diverse learners can thrive, managing classrooms effectively, and continuously adapting our teaching methods to meet the evolving needs of our students. Microlearning, with its concise, targeted approach to professional development, offers powerful strategies for achieving these goals. Drawing on my experiences as a principal at Kingsbury High School, Bennett High School and Getwell Elementary, where we leveraged microlearning to enhance teaching practices and address specific classroom challenges, this chapter explores practical ways to integrate microlearning into daily teaching routines, improve classroom management skills, and engage diverse learners.

Incorporating Microlearning into Daily Teaching Routines

Integrating microlearning into daily teaching routines begins with recognizing the natural intersections between professional development and classroom practice. As educators, the goal is to create a seamless flow where learning and application occur almost simultaneously. For instance, during my tenure at Kingsbury High School, we introduced short, daily professional development snippets focused on actionable teaching strategies. Teachers were encouraged to implement these strategies the same day, fostering an immediate connection between learning and teaching. This approach is supported by research from Bonsignore et al. (2020), who found that microlearning can significantly enhance teacher efficacy when integrated into regular teaching practices, as it promotes continuous reflection and adaptation.

We were an AVID School, so teachers had plenty of strategies to use and implement. At each faculty meeting a teacher shared a new strategy that they were implementing with fidelity, and they would demonstrate that strategy for the entire faculty. The expectation is that teachers would add that strategy to their toolbox as a resource and try the strategies immediately.

Using Microlearning for Classroom Management Skills

Effective classroom management is foundational to a productive learning environment. Microlearning modules that focus on specific aspects of classroom management, such as positive reinforcement techniques or strategies for de-escalating conflicts, can offer teachers quick, accessible solutions to common classroom challenges. In my role as President at New Teacher University, we developed a series of microlearning modules targeting these key areas. Teachers can access these modules on demand, allowing them to respond to classroom management issues with confidence and precision. New Teacher University offers strategies on how to implement Social Emotional Learning in your schools and classrooms, to strategies on how to establish workstations, including technology stations for Blended

Learning. This method aligns with findings from Smith & Harper (2021), who emphasize the importance of just-in-time professional development in building responsive and adaptable classroom management skills.

Strategies for Engaging Diverse Learners

Engaging diverse learners requires an understanding of varied learning styles, cultural backgrounds, and educational needs. Microlearning can facilitate this understanding by providing teachers with focused content on differentiation strategies, culturally responsive teaching, and techniques for supporting students with special needs. During my consultancy work with educational agencies, I advocated for the use of microlearning as a tool for enhancing teachers' abilities to engage diverse learners effectively. By offering concise, targeted insights and strategies, teachers can quickly adapt their lesson plans and teaching approaches to meet the needs of all students. This approach is supported by the work of Hernandez & Roberts (2018), who note that microlearning can play a crucial role in promoting inclusive education practices by enabling teachers to rapidly acquire and apply knowledge about diverse learner needs.

Finding short, powerful snippets of learning is better than not providing any learning at all. It is critical that when we provide microlearning opportunities that, they are purposeful and meaningful. Teachers do not like when administrators waste their time on general agenda items that could have been placed in a memo or email.

Incorporating microlearning into daily teaching routines, utilizing it to refine classroom management skills, and employing strategies to engage diverse learners are all effective ways to enhance teaching practices. By embracing microlearning, educators can ensure they are well-equipped to meet the challenges of modern teaching and provide all students with a high-quality education.

Chapter 5: The Impact of Microlearning on Teaching Performance

The pandemic followed by 'the great resignation' has left the field of education in a state of rapid evolution. This rapid evolution has caused educators to create innovative teaching methods to meet the challenges of their students. Many teachers discovered during the pandemic that they needed to figure out ways to engage students who were very comfortable with online learning. Among these, microlearning—a strategy that delivers content in small, manageable segments—has shown significant promise, especially in enhancing teaching performance. In this chapter, we will discuss the role of microlearning in building confidence and competence in new teachers and discuss the utility of microlearning in self-assessment practices.

Building Confidence and Competence in New Teachers

Microlearning stands out as an effective tool in the professional development of educators, particularly new teachers who are at the beginning of their careers. By breaking down complex information into digestible pieces, microlearning makes the learning process more accessible and less overwhelming. This approach aligns with cognitive load theory, which suggests that learners have a limited capacity for processing new information. Microlearning mitigates cognitive overload, facilitating a better understanding and retention of knowledge (Sweller, 1988).

When I was principal at Getwell Elementary School, microlearning principles were applied to professional development programs. By integrating short, focused training sessions on classroom management, instructional strategies, and student engagement, new teachers especially were able to rapidly assimilate essential skills. The microlearning occurred in faulty meetings, Professional Learning Communities PLCs, and collaborative planning times. This targeted approach not only built their confidence but also enhanced their competence, contributing to the school's transformation into one of the top-performing schools in the state. We were 90-90-90 school for two consecutive years. We were 97% title one, 95% proficient and advanced in ELA, and 94% proficient and advanced in Math.

Case Studies: Success Stories of Microlearning in Action

The application of microlearning extends beyond theoretical concepts, with numerous success stories underscoring its effectiveness. One such example is found in the initiatives implemented at Kingsbury High School, where my team and I led significant improvements in academic performance and school culture. Microlearning played a crucial role in professional development, where teachers engaged in brief, regular sessions focused on data analysis, differentiated instruction, and collaborative learning strategies. We were a pilot school for a rigorous English Curriculum that the district was interested in implementing. This curriculum was new for high schools because the books were consumable, and the teachers had some autonomy in how the instruction was delivered. Our English Department was a Rockstar in its implementation of the curriculum because of how we used the research on microlearning and the learning styles of millennials. This approach allowed for immediate application and reflection, leading to a 24% increase in ACT scores and 27% in graduation rates within two years. All the students were exposed to high-level rigorous text daily. Through microlearning and our teachers following the research, our scholars were the true winners.

Another case study involves the implementation of student-based budgeting systems, a complex process that requires careful planning and execution. Using microlearning, the staff at Kingsbury High School was gradually introduced to the new system, enabling them to understand and adapt to the changes effectively. Student Based Budgeting changed how we recruited, hired, onboarded, and assigned classes. Teachers had to participate in a number of capacity-building sessions in order for us to make the leap into designing our own team over having our standard allotment of staff given to us. This strategic use of microlearning facilitated a smoother transition, showcasing its versatility and impact on organizational change.

Reflecting on Practice: Microlearning for Self-Assessment

Microlearning also serves as a powerful tool for self-assessment and reflection, enabling educators to continuously evaluate and refine their teaching practices. By engaging with bite-sized learning materials, teachers can identify specific areas for improvement and immediately apply corrective measures. This iterative process fosters a culture of continuous learning and improvement.

For instance, the EPIC Gold Gain award received by my team and I at Getwell Elementary School highlights the importance of reflective practice in achieving excellence. Microlearning strategies, such as reviewing short case studies of effective teaching practices or reflecting on video clips of classroom interactions, were instrumental in identifying gaps and implementing targeted interventions. This not only led to significant academic gains but also helped in building a reflective practice among educators, where self-assessment became a routine part of their professional development. The reflective practice also revealed itself in the collaborative planning sessions and the PLCs. Teachers would regularly share their aha moments and pain points.

Microlearning is a pivotal strategy in enhancing teaching performance, particularly in building the confidence and competence of New Teachers, illustrating success through actionable case studies, and promoting a reflective practice among educators. The experiences of me and my teams in applying microlearning principles underscore the potential of this approach in creating impactful, sustainable changes in education. As research continues to support the efficacy of microlearning (Hug, 2005), its integration into professional development programs remains essential for fostering effective teaching and learning environments.

Chapter 6: Microlearning for Subject-Specific Expertise

As we work to refine educational strategies and enhance teaching effectiveness, microlearning is a powerful tool, not just for general pedagogical skills but also for gaining subject-specific expertise. This chapter delves into how microlearning can be adapted for different subject areas, its role in enhancing content knowledge and pedagogical skills, and the value of collaborative learning through the sharing of microlearning resources among peers.

Adapting Microlearning for Different Subject Areas

Microlearning's flexibility and adaptability make it particularly suitable for a wide range of subject areas. By delivering content in small, focused chunks, it allows for subject-specific nuances to be addressed effectively, catering to the unique needs of each discipline. For instance, at Kingsbury High School, microlearning was employed to tackle the diverse needs of a broad curriculum and a wide range of student needs. We had over 20 different languages spoken by our students at Kingsbury. Short, targeted sessions on topics such as mathematical problem-solving strategies or literary analysis techniques provided teachers with the specific insights needed to enhance their subject knowledge and instructional methods. The worst thing you can do as a principal is have the English Teachers in a Math Professional Development that doesn't have anything to do with language. Being able to customize your PD for teachers will save you a lot of side-eyes and OMGs in faculty meetings.

The adaptation of microlearning for different subjects involves tailoring content to the cognitive demands of the subject and the learning styles of the students. In science, for example, microlearning modules might focus on the 5 E's (Engage, Explore, Explain, Elaborate, and Evaluate), experimental procedures or concept visualization through animations and simulations. In contrast, for history or social studies, the emphasis could be on analyzing primary sources or engaging with interactive timelines.

Enhancing Content Knowledge and Pedagogical Skills

Microlearning significantly impacts both content knowledge and pedagogical skills by offering a dual focus on what is taught (content) and how it is taught (pedagogy). This is particularly important in subjects where educators may feel less confident or where there is a constant influx of new information. The microlearning approach adopted at Getwell Elementary School illustrates how breaking down complex educational strategies into manageable pieces can help teachers integrate new teaching methods into their practice, thereby enhancing their effectiveness across different subjects.

Research supports the effectiveness of this approach, with studies indicating that microlearning can significantly improve teachers' content knowledge and teaching skills, leading to better student outcomes

(Hodgins, 2002; Brame, 2016). By continuously updating their knowledge base and teaching strategies through microlearning, educators can stay current in their content and employ the most effective techniques in their classrooms.

Collaborative Learning: Sharing Microlearning Resources Among Peers

Collaboration among educators is a key component of professional growth and development. Microlearning facilitates this by making it easier to share resources and insights among peers. At Kingsbury High School, the sharing of microlearning resources contributed to a culture of collaborative learning. Teachers engaged in peer observations followed by micro-sessions where they discussed strategies, challenges, and successes. This not only broadened their repertoire of teaching strategies but also fostered a supportive professional community. Teachers were also given the opportunity to visit other schools to learn and come back and share in our PLCs and collaborative planning sessions.

The effectiveness of collaborative learning through microlearning is supported by Vygotsky's social learning theory, which posits that learning occurs within a social context and that interactions with peers play a critical role in the process of making meaning (Vygotsky, 1978). By exchanging microlearning resources and experiences, teachers can benefit from the collective knowledge and expertise of their colleagues, leading to improved teaching practices and student learning outcomes.

Microlearning offers a versatile and effective approach to developing subject-specific expertise among educators. Educators can significantly improve their teaching effectiveness by adapting microlearning strategies to cater to different subject areas, enhancing both content knowledge and pedagogical skills, and promoting collaborative learning through the sharing of resources. The experiences and strategies I share in this book exemplify the potential of microlearning to transform educational practices, providing a model for others seeking to leverage this approach in their own contexts.

Chapter 7: Leveraging Technology to Foster a Culture of Continuous Learning with Microlearning

Education as a practice has moved towards more personalized and efficient learning methods, and technology plays a major role in facilitating this shift. Technology is also major in fostering a culture of continuous learning among educators. Microlearning, with its flexible and learner-centered approach, serves as a pivotal tool in this endeavor. This chapter outlines strategies for encouraging a growth mindset among educators, building a supportive community of practice, and overcoming barriers to continuous professional development. This chapter also explores the various tools and platforms that support the creation and consumption of micro-learning content, the integration of social media and podcasts into professional learning, and the criteria for evaluating digital resources for microlearning.

The beginning of digital technology has revolutionized how microlearning content is created and consumed. Platforms such as Learning Management Systems (LMS), specialized microlearning apps, and online course providers offer versatile solutions for educators and learners alike. For example, platforms like Edmodo and Schoology are leveraged to distribute bite-sized learning materials that support both in-class instruction and independent study. These platforms facilitate the sharing of short videos, interactive quizzes, and other micro-content that enhance learning engagement and retention.

Furthermore, content creation tools such as Canva and Adobe Spark enable educators to easily design customized microlearning materials. The simplicity and accessibility of these tools empower teachers to rapidly produce high-quality, engaging content tailored to their students' needs.

Integrating Social Media and Podcasts into Professional Learning

Social media and podcasts are powerful platforms for microlearning, offering educators unprecedented opportunities for professional growth and networking. Social media platforms like X, YouTube, LinkedIn, and Facebook host vibrant educational communities where educators share insights, strategies, and microlearning resources.

Podcasts, on the other hand, provide an accessible way for educators to consume professional development content on-the-go. Educational podcasts cover a wide range of topics, from classroom management to subject-specific teaching strategies, and can be easily incorporated into a busy teacher's schedule. The convenience and flexibility of podcasts make them an ideal medium for microlearning, allowing educators to stay informed about the latest educational trends and research. The National New Teacher Support Podcast is a perfect tool for fostering a culture of continuous learning and collaboration among peers. The National New

Teacher Support Podcast is a podcast by award-winning teachers and administrators for teachers. The topics and the strategies are relevant to what teachers are experiencing now. All the strategies can be implemented with little to no resources. The National New Teacher Support Podcast specializes in strategies that can be implemented with will and skill.

Encouraging a Growth Mindset Among Educators

A growth mindset, the belief in the capacity to develop and improve one's abilities through dedication and hard work, is foundational to continuous learning. Microlearning can play a significant role in fostering this mindset by providing educators with accessible, manageable learning opportunities that demonstrate progress and encourage persistence. During Dr. Ross's tenure at Getwell Elementary, Bennett High, and Kingsbury High School, microlearning was utilized to instill a growth mindset by celebrating small wins and incremental progress in educators' professional development. This approach helped teachers see the value of ongoing learning and the potential for improvement in their teaching practices. As the teachers began to see the students benefitting from their wins, the implementation of microlearning became easier.

Research by Dweck (2006) supports the importance of a growth mindset in educational settings, showing that individuals who believe their abilities can be developed through hard work and effort are more likely to embrace challenges and persist in the face of setbacks. By integrating microlearning into professional development programs, educators can continuously update their skills and knowledge, reinforcing the belief in their own capacity to grow and adapt.

Building a Supportive Community of Practice

A community of practice, a group of individuals who share a concern or a passion for something they do and learn how to do it better as they interact regularly, is crucial for sustaining a culture of continuous learning. Microlearning facilitates the creation and maintenance of such communities by enabling the sharing of knowledge and experiences in a flexible and engaging manner. At Kingsbury High School, Dr. Ross fostered a supportive community of practice by using microlearning modules for collaborative training sessions. These sessions allowed educators to share insights, strategies, and feedback, thereby enhancing their collective expertise and fostering a sense of belonging and mutual support.

Wenger (1998) highlights the role of communities of practice in promoting learning and knowledge sharing within organizations. By participating in these communities, educators can leverage the collective wisdom of their peers, gaining new perspectives and ideas that can be applied to their teaching.

Overcoming Barriers to Continuous Professional Development

Despite the benefits of continuous professional development, educators often face barriers such as time constraints, lack of resources, and resistance to change. Microlearning addresses these challenges by offering flexible, bite-sized learning opportunities that can fit into busy schedules and require minimal resources. Furthermore, by delivering content in engaging and varied formats, microlearning can help overcome resistance to change by demonstrating the immediate value and applicability of new knowledge and skills.

Evaluating Digital Resources for Microlearning

With the vast array of digital resources available, it is crucial to evaluate their quality and effectiveness for microlearning. Criteria for evaluation include the relevance and accuracy of content, the credibility of the source, the engagement level of the material, and its alignment with learning objectives. Additionally, the ease of use and accessibility of the platform are important factors to consider, ensuring that all learners can benefit from the microlearning content.

The evaluation process can be informed by research and best practices in educational technology. For instance, the Technology Acceptance Model (TAM) suggests that perceived usefulness and ease of use are significant predictors of technology adoption (Davis, 1989). Educators like ourselves can apply such frameworks to assess and select the most appropriate digital resources for our microlearning initiatives.

By leveraging technology for microlearning, educators can enhance the learning experience, making it more engaging, flexible, and tailored to individual needs. The successful integration of tools and platforms for microlearning, as demonstrated through the initiatives led by my team and I, underscores the transformative potential of technology in education. As educators continue to explore and embrace these digital resources, they pave the way for more effective and personalized learning environments.

Sustaining Engagement and Motivation in Professional Learning

The key to long-term growth in any profession is sustained engagement and motivation. Microlearning contributes to this by providing quick, satisfying learning experiences that can easily be integrated into daily routines. My team and I at Getwell Elementary, Bennett High, and Kingsbury High School, leveraged microlearning as a way to stay engaged with new teaching methods, educational technologies, and pedagogical research without becoming overwhelmed. This approach aligns with the principles of self-determination theory, which emphasizes the importance of autonomy, competence, and relatedness in fostering motivation (Ryan & Deci, 2000). By allowing educators to choose what, when, and how they learn, microlearning supports these psychological needs, thereby enhancing motivation and engagement over the long term. The Instructional Leadership Team uses data to establish what we need to improve on, and from those choices, the team will be able to choose. It is important that the PD and method of learning are relevant to the school's goals, vision , and mission.

Setting Milestones and Celebrating Achievements

Effective long-term planning in professional learning involves setting clear milestones and taking the time to celebrate achievements, no matter how small. These practices not only provide a sense of progress and accomplishment but also serve to reinforce motivation and commitment to continuous improvement. The implementation of microlearning strategies should include regular reviews of educators' progress against predefined milestones, such as the integration of a new teaching strategy or the completion of a series of microlearning modules on a specific topic. Celebrating these achievements, perhaps through recognition in staff meetings or via digital badges, helps to build a culture of appreciation and success. The goal-setting theory of motivation supports this approach, suggesting that specific and challenging goals lead to higher performance (Locke & Latham, 2002).

Future-Proofing Your Teaching Career with Ongoing Microlearning

With rapid technological advancements and shifting educational priorities, future-proofing one's teaching career is more important than ever. Microlearning facilitates this by enabling educators to continuously update their skills and knowledge in alignment with the latest educational trends and research findings. This not only benefits individual teachers but also contributes to the overall quality of education provided to students.

The concept of lifelong learning is crucial in this regard, with microlearning serving as a key enabler by providing flexible, just-in-time learning opportunities that can be pursued throughout an educator's career (Jarvis, 2004). By adopting a proactive approach to professional development, educators can adapt to changes, embrace new challenges, and ensure their teaching remains impactful and relevant.

Planning for long-term growth with microlearning involves sustaining engagement and motivation, setting and celebrating milestones, and continuously updating one's skills and knowledge to future-proof their teaching career. My experiences and leadership offer valuable insights into how these strategies can be effectively implemented, underscoring the importance of microlearning in supporting educators' professional development and long-term success.

Chapter 8: Getting Started and Planning for Long-Term Growth with Microlearning

Embarking on a journey of professional development through microlearning can seem daunting at first, but with the right approach, it becomes an enriching path to continuous improvement and personal growth. With its flexible and targeted approach, micro learning offers a promising avenue for educators to engage in continuous professional development, ensuring their teaching practices remain relevant and effective. This chapter will explore strategies for sustaining engagement and motivation in professional learning, setting milestones and celebrating achievements, and future-proofing teaching careers through ongoing microlearning. Finally, this chapter provides a roadmap for launching your microlearning initiative, staying committed to your professional development plan, and leveraging resources and tools to maximize your microlearning journey, incorporating the experiences and strategies that I have learned over the years.

First Steps: Launching Your Microlearning Initiative

The initial step in starting a microlearning initiative is to identify your learning goals and objectives. Reflect on the areas of your professional practice where you seek improvement or new knowledge. Once goals are established, select a microlearning format that best suits your learning style and schedule. This could range from short articles and videos to interactive quizzes and infographics. The key is to choose content that is engaging and can be easily integrated into your daily routine. Establishing a regular learning schedule, even if it's just a few minutes each day, can help build a habit and ensure consistent progress.

Tips for Staying Committed to Your Professional Development Plan

Staying committed to a professional development plan requires motivation, discipline, and a supportive network. Here are some strategies to maintain your commitment:

1. **Set Realistic Goals**: Break your long-term objectives into smaller, manageable milestones. Celebrate achieving these milestones to maintain motivation.

2. **Leverage Peer Support**: Engage with colleagues or an online community of professionals who share your interest in microlearning. Collaborative planning teams exemplify the power of a supportive community in sustaining engagement and fostering a culture of continuous learning.

3. **Reflect and Adjust**: Regularly assess your progress towards your goals. Be prepared to adjust your learning plan based on what is or isn't working, ensuring it remains aligned with your evolving professional needs. Looking at student data and student work is a

great way to assess your progress.

4. **Incorporate Learning into Daily Activities**: Find ways to integrate microlearning into your everyday tasks. For example, listening to educational podcasts during commutes or reading articles during breaks can make learning more accessible and less time-consuming.

Resources and Tools to Jumpstart Your Microlearning Journey

A wealth of resources and tools are available to support your microlearning initiative. Here are some to consider:

- **Learning Management Systems (LMS)**: Platforms like Edmodo, Moodle, and Schoology offer courses and content in micro-formats, allowing for personalized learning experiences.
- **Content Creation Tools**: Tools such as Canva, Adobe Spark, and Powtoon enable educators to create their own microlearning materials, tailoring content to specific needs and interests.
- **Online Course Providers**: Websites like Coursera, Udemy, and Khan Academy provide access to a wide range of subjects, offering short courses and modules that fit the microlearning model.
- **Podcasts and Blogs**: A plethora of educational podcasts and blogs cover virtually every topic imaginable, offering insights and knowledge in concise formats.

In launching your microlearning initiative, remember the importance of flexibility, persistence, and a willingness to adapt. My journey demonstrates the transformative potential of microlearning in achieving professional excellence and impacting educational outcomes. By taking the first steps, committing to your development, and utilizing available resources and tools, you set the stage for a fulfilling and impactful professional growth journey.

Embarking on this path not only enhances your skills and knowledge but also contributes to the broader educational community, fostering a culture of continuous learning and improvement.

Appendices

Recommended Microlearning Platforms and Tools for Teachers and Educators

Microlearning Platforms

1. **Edmodo**: A versatile platform that allows teachers to share bite-sized learning resources, create quizzes, and foster collaboration among students.

2. **Kahoot!**: Known for creating engaging quizzes and interactive challenges, Kahoot! makes learning fun and can be used for quick knowledge checks.

3. **Quizlet**: Offers study tools and flashcards that are great for microlearning. Teachers can create their own sets or use existing ones to support learning.

4. **Socrative**: A classroom app for formative assessments and instant feedback, Socrative supports microlearning through quizzes, quick questions, and exit tickets.

5. **Ted-Ed**: Provides access to educational videos on a wide range of topics, which can be used as microlearning content in various subjects.

6. **Coursera and Udemy**: While known for more extensive courses, they also offer shorter courses and modules perfect for

microlearning in professional development.

7. **Duolingo**: Ideal for language teachers, Duolingo offers bite-sized lessons that are effective for both personal learning and classroom instruction.

8. **BrainPOP**: Features animated educational videos, quizzes, and games across subjects, suitable for younger learners and as a microlearning tool for teachers.

Content Creation Tools

1. **Canva**: User-friendly graphic design tool for creating infographics, presentations, and educational materials that can serve as microlearning content.

2. **Adobe Spark**: Enables teachers to create short videos, web pages, and graphics to support learning in engaging and visually appealing formats.

3. **Powtoon**: A tool for creating animated videos and presentations, perfect for explaining complex concepts in simple, engaging ways.

4. **Anchor**: A podcasting tool that allows educators to easily create, distribute, and manage podcasts, which can be a great medium for microlearning.

5. **Flipgrid**: A social learning platform that lets teachers and students create and share short video responses to prompts, facilitating discussion and reflection.

6. **Loom**: With Loom, educators can create quick video lessons or tutorials by recording their screen, voice, and face, making complex ideas easier to understand.

Collaboration and Sharing Tools

1. **Google Classroom**: Integrates seamlessly with other Google services, making it easy to share educational materials, assign tasks, and communicate with students.

2. **Microsoft Teams for Education**: Provides a platform for virtual classes, assignments, and collaborative projects, supporting a continuous learning environment.

3. **Slack**: While not traditionally educational, Slack can be used to create communities of practice among educators for sharing resources and best practices.

Each of these platforms and tools offers unique features to support the implementation of microlearning in educational settings. By selecting the resources that best fit their needs and teaching style, educators can enhance their instructional strategies and contribute to a culture of continuous learning among their students.

Directory of Microlearning Resources for Various Subjects

A directory of microlearning resources tailored to various subjects can significantly aid educators in finding targeted, high-quality content for their teaching and professional development needs. Below is a structured guide to microlearning resources across a range of subjects, designed to provide quick access to tools and platforms that support concise, impactful learning experiences.

Mathematics

- **Khan Academy**: Offers comprehensive lessons on a wide range of math topics, from basic arithmetic to advanced calculus, in easily digestible video formats.
- **IXL**: Provides interactive math practice for all grade levels, with questions that adapt to students' skill levels for personalized learning experiences.

Science

- **PhET Interactive Simulations**: Hosted by the University of Colorado Boulder, offering engaging, interactive simulations in physics, chemistry, biology, and earth science.
- **NASA's Teachable Moments**: Brings the latest space and science discoveries to the classroom through bite-sized educational resources.

Language Arts

- **NoRedInk**: Builds stronger writers through interest-based curriculum, adaptive exercises, and actionable data, focusing on grammar and writing.

- **ReadTheory**: Personalized reading comprehension exercises for students in all grade levels, adapting to student performance to help improve reading skills over time.

Social Studies

- **TED-Ed's History & Social Studies**: Curates educational videos that cover significant historical events and social studies concepts, perfect for microlearning.

- **iCivics**: Founded by Justice Sandra Day O'Connor, it provides engaging games and resources to teach students about civics and government.

Languages

- **Duolingo**: Offers bite-sized lessons in over 30 languages, using gamification to make language learning engaging and effective.

- **Memrise**: Uses spaced repetition and mnemonic techniques to help learn vocabulary and phrases in various languages through short, interactive lessons.

Professional Development for Educators

- **EdSurge**: Provides news, research, and tools for digital learning, perfect for educators looking to stay on top of educational technology trends.

- **Coursera and Udemy**: While they offer comprehensive courses, many are segmented into shorter lessons that are ideal for educators seeking to expand their pedagogical or subject-specific knowledge.

Arts & Music

- **SmartHistory**: A leading resource for art history, offering an extensive collection of videos and essays on art and cultural heritage from around the world.
- **Yousician**: An interactive music service to learn guitar, piano, bass, ukulele, or singing with microlessons that provide instant feedback.

Computer Science & Technology

- **Codecademy**: Offers interactive coding lessons in various programming languages, making it easy to build coding skills in digestible increments.
- **Cybrary**: Focuses on cybersecurity and IT skills, providing micro-courses on a range of topics from ethical hacking to information security.

Health & Physical Education

- **Sworkit**: Provides customizable workout plans that can fit into any schedule, supporting physical education with short, effective exercise routines.
- **Healthline**: Offers a wealth of articles and videos on various health topics, ideal for quick lessons or updates on health and wellness

trends.

Special Education

- **Understood**: Offers resources and short guides for educators and parents of students with learning and attention issues, providing strategies and support for inclusive education.
- **PBS LearningMedia Special Education**: A collection of resources designed to support the diverse needs of students, with materials ranging from videos to interactive lessons.

This directory aims to serve as a starting point for educators exploring microlearning resources across different subjects. By integrating these tools into their teaching strategies, educators can enhance engagement, personalize learning experiences, and support students' educational journeys in a flexible and impactful manner.

Personalized Microlearning Plan

A personalized microlearning plan involves identifying specific learning goals, selecting relevant resources, scheduling learning sessions, and evaluating progress. Below is a template that educators, professionals, or anyone interested in personal or professional development can adapt to their learning needs. This template is designed to be flexible, allowing for adjustments based on individual preferences and goals.

Personalized Microlearning Plan Template

This template serves as a comprehensive guide to structuring a microlearning plan that is personalized to your learning objectives and preferences. It emphasizes the importance of setting clear goals, utilizing diverse resources, maintaining a consistent learning schedule, and regularly assessing progress to ensure sustained engagement and successful outcomes.

References

- Dweck, C. (2006). *Mindset: The New Psychology of Success*. Ballantine Books.
- Wenger, E. (1998). *Communities of Practice: Learning, Meaning, and Identity*. Cambridge University Press.
- Ryan, R.M., & Deci, E.L. (2000). Self-determination theory and the facilitation of intrinsic motivation, social development, and well-being. *American Psychologist*, 55(1), 68-78.
- Locke, E.A., & Latham, G.P. (2002). Building a practically useful theory of goal setting and task motivation: A 35-year odyssey. *American Psychologist*, 57(9), 705-717.
- Jarvis, P. (2004). *Adult Education and Lifelong Learning: Theory and Practice*. Routledge Falmer.
- Davis, F. D. (1989). Perceived Usefulness, Perceived Ease of Use, and User Acceptance of Information Technology. MIS Quarterly, 13(3), 319-340.
- Platforms such as Edmodo, Schoology, Canva, and Adobe Spark are widely recognized in the educational technology community for their utility in creating and distributing microlearning content.
- The use of social media and podcasts for professional learning is supported by literature on informal learning and professional

development through digital platforms (Greenhow, C., & Lewin, C., 2016).

- Hodgins, H. (2002). The future of learning objects. In Proceedings of the 2002 eTEE Conference.
- Brame, C.J. (2016). Effective Educational Videos: Principles and Guidelines for Maximizing Student Learning from Video Content. *CBE—Life Sciences Education*, 15(4), es6.
- Vygotsky, L.S. (1978). *Mind in Society: The Development of Higher Psychological Processes*. Harvard University Press.
- Bonsignore, R., et al. (2020). Microlearning in Teacher Professional Development: Engagement and Application in the Classroom. *Journal of Teacher Education*, 71(4), 472-489.
- Hernandez, D., & Roberts, T. (2018). Culturally Responsive Teaching and Microlearning: Supporting Teachers in Inclusive Education. *International Journal of Inclusive Education*, 22(7), 803-817.
- Smith, L., & Harper, N. (2021). Just-in-Time Professional Development: Microlearning for Quick Classroom Application. *Teaching and Teacher Education*, 39(2), 101-113.
- Garet, M. S., Porter, A. C., Desimone, L., Birman, B. F., & Yoon, K. S. (2011). What makes professional development effective? Results from a national sample of teachers. *American Educational Research Journal*, 38(4), 915-945.

- Johnson, C., & Bartlett, J. (2019). The use of podcasting to enhance professional development for teachers. *TechTrends*, 63(1), 100-105.
- Mehta, R., & Aguilera, E. (2020). A critical approach to humanizing pedagogies in online teaching and learning. *International Journal of Information and Learning Technology*, 37(3), 109-120.
- Sweller, J. (1988). Cognitive Load During Problem Solving: Effects on Learning. *Cognitive Science*, 12(2), 257-285.
- Hug, T. (2005). Microlearning: A new pedagogical challenge. *Introduction to Microlearning*, 1-12.

Made in the USA
Monee, IL
04 November 2026

67383935R00036